BEES

by Joyce Markovics

NORWOOD House Press

For more information about Norwood House Press, please visit our website at: www.norwoodhousepress.com or call 866-565-2900.

Book Designer: Ed Morgan
Editorial and Production: Bowerbird Books

Photo Credits: freepik.com, cover; freepik.com, title page; freepik.com, 5; © iStock.com/akiyoko, 6 bottom; © Dr. Stephen L. Buchmann, 6 top; freepik.com, 7; Egor Kamelev/Pexels, 8; freepik.com, 9; USGS/Wikimedia Commons, 10; USDA ARS/Wikimedia Commons, 11 top; freepik.com, 11 bottom; freepik.com, 12; Courtesy Damien Tupinier/Unsplash.com, 13 top; freepik.com, 13 bottom; Vipin Baliga/flickr, 14; Waugsberg/Wikimedia Commons, 15 top; Jon BeesinFrance/flickr, 15 bottom; Unsplash.com, 16–17; freepik.com, 18; Unsplash.com, 19; freepik.com, 20 left; freepik.com, 20 right; Susi Rosenberg/flickr, 21; freepik.com, 22; freepik.com, 23; freepik.com, 24; USGS/flickr, 25; Gilles San Martin/flickr, 25; Unsplash.com, 27; freepik.com, 29.

Copyright © 2023 Norwood House Press

Hardcover ISBN: 978-1-68450-766-5
Paperback ISBN: 978-1-68404-777-2

All rights reserved. No part of this book may be reproduced or utilized in any form or by any means without written permission from the publisher.

Library of Congress Cataloging-in-Publication Data

Names: Markovics, Joyce L., author.
Title: Bees / by Joyce Markovics.
Description: Chicago : Norwood House Press, [2023] | Series: Nature's friends | Includes bibliographical references and index. | Audience: Grades 2-3
Identifiers: LCCN 2021057764 (print) | LCCN 2021057765 (ebook) | ISBN 9781684507665 (hardcover) | ISBN 9781684047772 (paperback) | ISBN 9781684047833 (ebook)
Subjects: LCSH: Bees--Juvenile literature.
Classification: LCC QL565.2 .M37 2023 (print) | LCC QL565.2 (ebook) | DDC 595.79/9--dc23/eng/20211217
LC record available at https://lccn.loc.gov/2021057764
LC ebook record available at https://lccn.loc.gov/2021057765

353N—082022

Manufactured in the United States of America in North Mankato, Minnesota.

CONTENTS

Being a Bee	**4**
Pollinators	**8**
Cool Colony	**12**
Sweet Like Honey	**16**
Beekeeping	**18**
Bees at Risk	**22**
Pests and Poisons	**24**
A World without Bees	**26**
Nurture Nature: Plant a Bee Garden!	28
Glossary	30
For More Information	31
Index	32
About the Author	32

BEING A BEE

Zoom! A worker honeybee flies above a garden. The sun shines like a golden ball in the sky. Below the bee is a rainbow of flowers. The most colorful ones stand out like **neon** signs. They tell the bee to come closer. She spots a pink rose and buzzes around it. The bee knows the blossom is rich in nectar. With her straw-like tongue, she tastes the sweet liquid.

All worker honeybees are female. A worker will share a sample of the nectar she's collected with other workers in her hive.

Then the honeybee makes an unexpected move. Instead of feeding, she flies back to her hive. Once inside the hive, she starts dancing! She wiggles her **abdomen** and walks around. Her wings buzz and twitch. Her dancing directs other worker bees to the rich food source she's found.

A honeybee

> Why do bees buzz? The buzzing sound is made by the rapid movement of their wings.

Bees are a type of insect. They have six legs and three main body parts. Bees come in all shapes and sizes. There are 20,000 bee **species** worldwide. They range in size from the tiny *Perdita minima* to the paperclip-sized carpenter bee. Bees can be found in mountains, fields, and deserts. In fact, they live anywhere that flowers bloom!

A *Perdita minima*—the world's smallest bee—on a carpenter bee's head

A carpenter bee

Honeybees are not native to, or naturally found in, the United States. When Europeans came to America around 1600, they brought honeybees with them. The bees soon escaped from their hives. They set up large groups called colonies in trees or underground. Since then, honeybees and native bees have worked together to pollinate flowers.

People have been keeping honeybees for many hundreds of years. Beekeepers often wear hooded suits. The suits protect them from getting stung by the bees.

POLLINATORS

What is pollination? When a bee lands on a flower, it gathers pollen. The tiny yellow grains stick to the bee's fuzzy body. The bee also collects pollen using long hairs on its legs. When the bee flies to another flower, some of the pollen rubs off on the new bloom. After this occurs, the plant can make seeds. Over time, the seeds grow into new plants!

A honeybee's fuzzy body fully covered in pollen

Bees are tiny treasures. They help pollinate thousands of different plants. These plants include eighty percent of flowers. Bees also pollinate about seventy-five percent of fruit and nut trees, such as apple and almond. They pollinate vegetables too. Without bees, there would be far fewer yummy foods for people to eat.

Bees and other animals that transfer pollen from one flower to another are called pollinators.

The blueberry bee is one of the hardest-working pollinators. And it has a favorite food—blueberry flowers! A single worker can visit 60,000 blueberry flowers in its life. When a blueberry bee lands on a bloom, she **vibrates** her body. She does this to shake the pollen out of the flower. In a few weeks, she'll pollinate enough flowers to create 6,000 blueberries. That's a lot of blueberry pie!

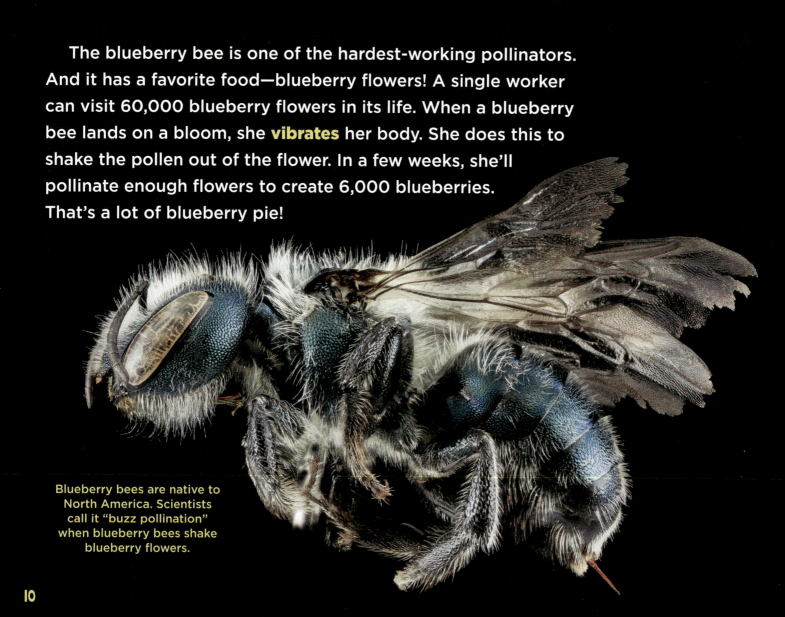

Blueberry bees are native to North America. Scientists call it "buzz pollination" when blueberry bees shake blueberry flowers.

Squash bees pollinate squash, pumpkin, and melon flowers. They're about the same size and color as honeybees. They start work early. At dawn, squash flowers open. That's when squash bees make their morning move. Once they find a flower, they plunge their bodies inside. As they do, the bees gather loads of pollen and nectar.

Squash bees will sleep inside flowers after they feed!

Squash bees live underground, below the plants they pollinate. Have you ever picked your own pumpkin? If so, there were probably nests of baby squash bees under your feet!

COOL COLONY

Most bees, such as blueberry bees, live alone. Others, like honeybees, live in big colonies. Honeybee colonies can include up to 80,000 bees! A honeybee colony includes three types of adult bees. These are workers, **drones**, and a queen. The workers collect food, defend the colony, help raise the young, and care for the queen. Drones are male bees that **mate** with the queen. Their only job is to help the queen produce young.

Honeybees can't survive without help from every single member of the colony.

The queen bee is bigger than all the other bees. She can live up to five years. A worker only lives about five weeks!

Each colony has one queen. Her job is laying eggs. She can lay about 1,500 a day in spring and summer. In her lifetime, she may produce up to one million eggs. The queen also makes **pheromones**. These chemical scents allow her to "talk" with the other bees. They tell the workers what to do and when to do it. Pheromones help all the bees in the colony function as one unit.

Honeybees can sting. However, they rarely do unless they're defending the hive.

Honeybee eggs inside their cells

The queen lays her eggs in the hive. The inside of the hive is made up of a structure called a honeycomb. Worker bees build the honeycomb out of beeswax. The wax is made by **glands** in the bees' bodies. Then the bees chew the wax. They shape it into small, six-sided rooms, or cells. All the cells are exactly the same size.

The queen deposits one egg inside each cell. Each egg is about the size of a grain of rice. After three days, the eggs hatch into wormlike **larvae**. Special workers care for the baby bees. They feed them and keep them warm and safe. About nine days after the larvae hatch, the workers seal them into their cells. Inside the cells, the young bees, now called **pupae**, transform. Within a few weeks, they become adult insects!

These honeybee young have been removed from their cells. They are shown at different stages of development, from larvae to pupae to adults (left to right).

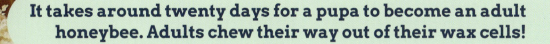

It takes around twenty days for a pupa to become an adult honeybee. Adults chew their way out of their wax cells!

15

SWEET LIKE HONEY

Honeybees are best known for one thing—honey! To make this sweet, sticky treat, workers collect lots of nectar. They fill up their stomachs with as much nectar as they can carry. Then they fly back to the colony.

The bees spit up the nectar into the other workers' mouths. The honey-making workers pass the nectar from bee to bee. This reduces the amount of **moisture** in the nectar. The nectar becomes thick and gooey. Sometimes, the nectar is stored in cells. Workers may fan the nectar with their wings. This helps it **evaporate** and thicken. Finally, the honey is sealed with wax in cells for future use.

One bee colony can make 100 pounds (45 kg) of honey per year!

Honey, as well as pollen, is used as food for the colony.

BEEKEEPING

Many people raise bees for honey and beeswax—or just for fun! They keep bees in human-made hives. Most hives look like a stack of wooden boxes. Inside the boxes are frames where the bees build their combs. Beekeepers do a lot to care for their bees. They make sure they have enough food. They also protect their bees from illnesses and **predators**, such as bears.

A beekeeper examines bees on a frame for any signs of illness.

Some people raise bees in cities. They set up hives in gardens, on roofs, or even in their homes. One New York City beekeeper keeps his bees in a sealed room in his apartment. The insects come and go through a small window! Amazingly, beekeepers raise more than half of all the bees in the United States.

This city beekeeper uses a special hive called a B-BOX. It allows her to harvest honey without coming into contact with any bees.

Beeswax has many uses. People use it to make candles, makeup, creams, and even furniture polish!

Some people raise honeybees for business. One of the biggest bee businesses is almonds. Most of the world's almond trees are in California. Honeybees are superb at pollinating these trees. However, there aren't enough local bees to do the job. So beekeepers bring their beehives to the almond trees!

Beekeepers stack their hives on huge trucks. Some drive them all the way across the country. Once they arrive in the almond **groves**, the keepers unload their bees. The buzzing worker bees zip out of their hives and get to work. The bees will pollinate billions of the blooms in a matter of weeks.

However, not all is well with bees. In fact, the small insects are facing a giant problem.

An almond tree orchard in California

A honeybee about to pollinate an almond flower

California is home to eighty percent of the world's almond trees.

BEES AT RISK

In recent years, bees have been dying off. More than thirty percent of honey bees are lost every year. Why are they dying? There are lots of reasons, say bee experts. For one, bees need many flowers to survive. Yet people are destroying the places where bees live and feed. They're plowing up wildflower meadows to make room for farmland and buildings. Fewer flowers mean less food for hungry bees.

Climate change may also be to blame for falling bee numbers. Extreme weather could affect when plants flower. Bees might miss peak pollination time. In parts of the world where bees are **scarce**, farmers pollinate their crops by hand. They brush blossoms with small paintbrushes dipped in pollen! This slow process can take weeks.

A farmer pollinates a strawberry flower with a paintbrush.

PESTS AND POISONS

Honeybees are also facing two other big problems: **parasites** and **pesticides**. Both cause a large number of bee deaths. The main parasite that attacks bees is the varroa (VAHR-oh-uh) **mite**. The tiny crab-like animal latches onto a honeybee's body. Then the mite releases chemicals.

The chemicals break down the honeybee's body. They turn the bee's flesh into a creamy liquid. Then the killer mite slurps up the bee's insides! This can cause the bee to get sick or die. Pesticides are deadly to bees as well. Farmers spray these poisons on their crops to destroy pests. Yet the poisons can also weaken or kill bees.

The brown dot on the bee's body is the parasitic varroa mite.

A female varroa mite on the head of a bee pupa

Experts have found tiny amounts of up to 14 different pesticides in some bee colonies.

A WORLD WITHOUT BEES

What if there were no bees? To start, people would lose eighty to one hundred different kinds of food crops. Without these crops, there would be less food for people. It's thought that grocery stores would have half as much food. Even worse, all the plants that bees pollinate could die. The animals that feed on these plants might **perish** too.

A world without bees could mean a world without many foods and animals. So what can we do to help? Bee expert Sammy Ramsey says, "plant tons of flowers in your front and backyard." Leave dandelions alone, and don't use pesticides. "I think bees actually really love being appreciated," says Sammy. With enough flowers, bees have a good chance of making a comeback.

Beehives in a person's backyard

Bees also pollinate grapes, coffee, avocados, and oranges. Some experts say the crops that bees pollinate feed ninety percent of the world!

27

PLANT A BEE GARDEN!

More and more bees are dying every day. However, protecting bees is easier than you think. One simple way is to plant a bee garden!

- Ask an adult to help you get a big flower pot with drainage holes.
- With an adult's help, buy some seeds at a garden store. Flowers such as marigolds, sunflowers, and cosmos are good options.
- Fill the pot with soil. Then plant a few seeds. Use your finger to push the seeds into the soil.
- Place the pot outside in a sunny spot. Lightly water the seeds every day.
- In a few weeks, the plants will grow. Eventually, they will bloom and attract bees.

WATCH THE BEES ENJOYING THEIR FLOWER FEAST.
HAPPY BUZZING!

GLOSSARY

abdomen (AB-duh-muhn): the back part of an insect's body.

climate change (KLYE-mit CHAYNJ): the warming of Earth's air and oceans due to environmental changes, such as a buildup of greenhouse gases that trap the sun's heat in Earth's atmosphere.

drones (DROHNZ): male bees whose job it is to mate with the queen.

evaporate (i-VAP-uh-rayt): to dry up; to change from a liquid into a gas.

glands (GLANDS): body parts that secrete chemicals.

groves (GROHVS): orchards of fruit-bearing trees.

larvae (LAR-vee): the wormlike form of many kinds of young insects; the singular form is larva.

mate (MAYT): come together to have young.

mite (MAHYT): a small spider-like animal, related to ticks, that often feeds on other animals.

moisture (MOIS-chur): very small drops of water or other liquid.

neon (NEE-on): a type of sign that contains a glowing gas.

parasites (PA-ruh-sites): living creatures that get food by living on or in another plant or animal.

perish (PAIR-ish): to die.

pesticides (PESS-tuh-sidz): chemicals that kill insects and other pests that damage crops.

pheromones (FAIR-uh-mohnz): chemicals with a scent produced by animals to send a message to other animals.

predators (PRED-uh-turz): animals that hunt and kill other animals for food.

pupae (PYOO-pee): young insects in a form between larvae and adults.

scarce (SKAIRS): hard to find.

species (SPEE-sheez): types of animals or plants.

vibrates (VYE-brates): moves back and forth quickly.

FOR MORE INFORMATION

Books

Fleming, Candace. *Honeybee: The Busy Life of Apis Mellifera*. New York, NY: Neal Porter Books, 2020.
This book explores the life of a worker honeybee.

Milner, Charlotte. *The Bee Book*. New York, NY: DK Children, 2018.
Read this book to learn amazing facts about bees.

Socha, Piotr. *Bees: A Honeyed History*. New York, NY: Abrams Books for Young Readers, 2017.
Readers will learn about the history of beekeeping.

Websites

**Brooklyn Botanic Garden: Native Bees
(https://www.bbg.org/gardening/article/native_bees)**
Readers can learn about different kinds of bees.

**San Diego Zoo
(https://animals.sandiegozoo.org/animals/bee)**
Find out bee basics, including habitat and diet.

**UC Berkeley: Urban Bee Lab
(http://www.helpabee.org)**
Explore city bee gardens.

INDEX

almond trees, 20, 21
beekeeping, 18, 19, 20, 21
beeswax, 14, 18, 19
blueberry bees, 10, 12
cells, 14, 15, 16
climate change, 23
colony, 12, 13, 16, 17
crops, 23, 25, 26, 27
drones, 12
eggs, 13, 14, 15

flowers, 4, 6, 7, 8, 9, 10, 11, 22, 23, 26
food, 5, 9, 10, 12, 16, 17, 18, 22, 26
hives, 5, 7, 13, 14, 18, 19, 20
honey, 16, 17, 18, 19
honeycomb, 14
larvae, 15
nectar, 4, 11, 16
parasites, 24, 25

pesticides, 24, 25, 26
pollen, 8, 9, 10, 11, 17, 23
pollination, 7, 8, 9, 10, 11, 20, 23, 26, 27
predators, 18
pupae, 15
queen, 12, 13, 14, 15
squash bees, 11
worker bees, 4, 5, 10, 12, 13, 14, 15, 16, 20

ABOUT THE AUTHOR

Joyce Markovics has written hundreds of books for kids. She's wild about bees and insects of all kinds. Joyce lives in an old, creaky house along the Hudson River. She hopes the readers of this book will take action—in small and big ways—to protect nature, one of our greatest gifts. Joyce dedicates this book to beekeepers everywhere.